Siri Matematik BurujKids

BurujKids Maths Series

HAFALAN SIFIR MUDAH /

SIMPLE TIMES TABLE MEMORIZATION

(Terdapat dalam dwi-bahasa / *Available in bilingual*)

PARTRIDGE PUBLISHING SINGAPORE
2019

To order additional copies of this book, contact
Toll Free 800 101 2657 (Singapore)
Toll Free 1 800 81 7340 (Malaysia)
www.partridgepublishing.com/singapore
orders.singapore@partridgepublishing.com

01/16/2020

PARTRIDGE

KANDUNGAN / TABLE OF CONTENTS

Stail Satu Satu Sifar (110) / One-One-Zero Style (110)

(Tangan Kanan / Right Hand)

Scan me

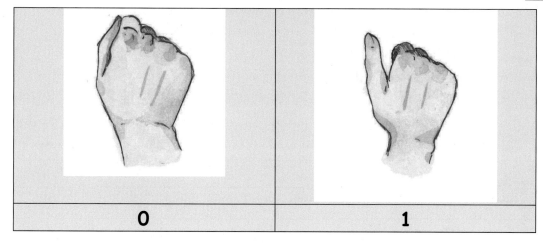

| 0 | 1 |

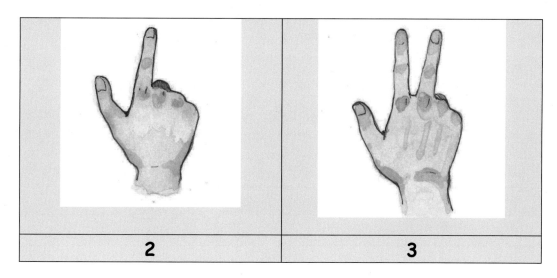

| 2 | 3 |

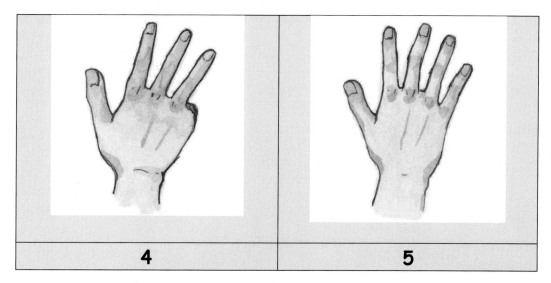

| 4 | 5 |

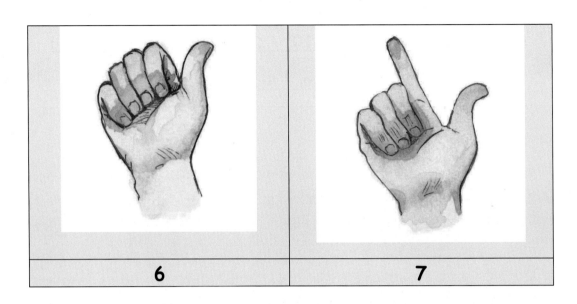

| 6 | 7 |

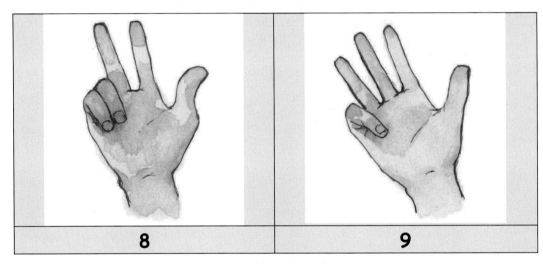

| 8 | 9 |

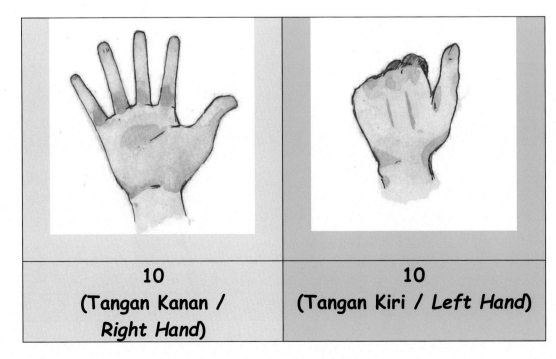

| 10 (Tangan Kanan / Right Hand) | 10 (Tangan Kiri / Left Hand) |

Stail Satu Satu Sifar (110) / One One Zero Style (110)

(Tangan Kiri / Left hand)

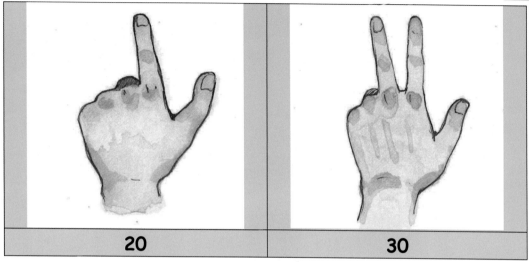

20	30

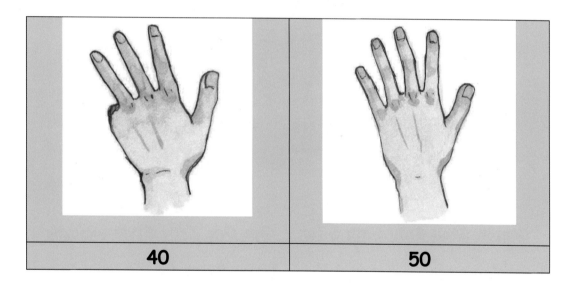

40	50

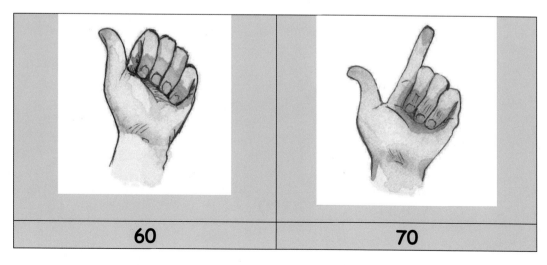

60	70

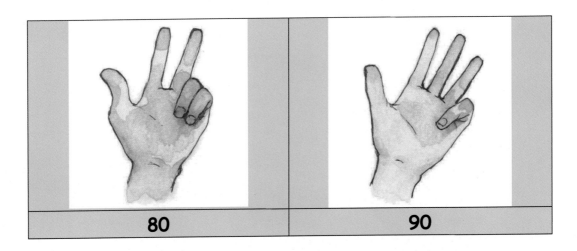

80   90

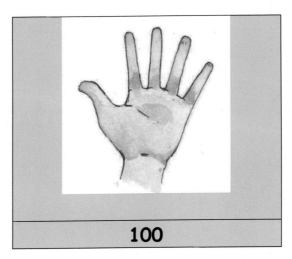

100

Contoh Stail Satu Satu Sifar / One One Zero Style example

(Guna tangan Kanan & Kiri / Using Right & Left hands)

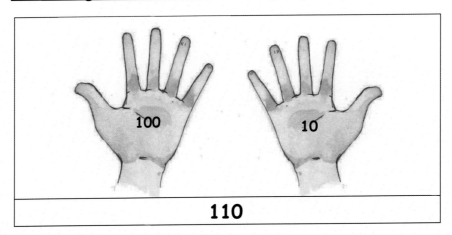

100   10

110

Penerangan untuk Stail Satu Satu Sifar /

Explanation on the One-One-Zero Style

Sifir / Times	Menggunakan kedua-dua tangan / Using both hands	
	Tangan kiri / Left hand	Tangan kanan / Right hand
1	-	+ 1 jari / + 1 finger
2	-	+ 2 jari / + 2 fingers
5	-	+ 5 jari / + 5 fingers
10	+ 1 jari / + 1 finger	-
11	+ 1 jari / + 1 finger	+ 1 jari / + 1 finger
12	+ 1 jari / + 1 finger	+ 2 jari / + 2 fingers

Sifir / Times	Tangan kanan sahaja / Using right hand only
9	- 1 jari / - 1 finger
8	- 2 jari / - 2 fingers
6	+ 1 jari & terbalikkan tangan / + 1 finger & turn hand
7	+ 2 jari & terbalikkan tangan / + 2 fingers & turn hand
4	- 1 jari & terbalikkan tangan / - 1 finger & turn hand
3	- 2 jari & terbalikkan tangan / - 2 fingers & turn hand

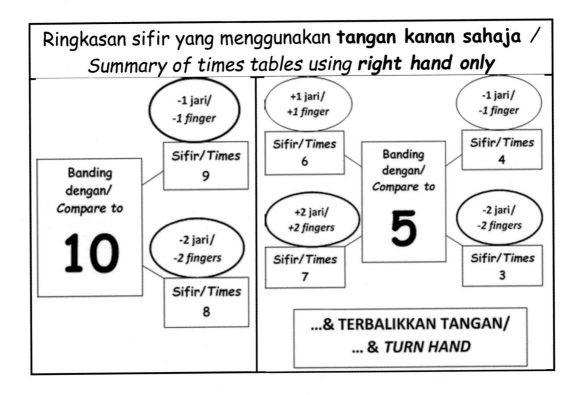

Ringkasan sifir yang menggunakan **tangan kanan sahaja** / *Summary of times tables using* **right hand only**

Ada beberapa peraturan apabila menggunakan **tangan kanan sahaja**... / *There are a few rules to follow when using* **only the right hand**...

1. Apabila bilangan jari berkurangan, tambahkan satu nilai puluh / *When the number of fingers is decreasing, add one tens.*

2. Apabila bilangan jari bertambah, tetap pada nilai puluh yang sama / *When the number of fingers is increasing, stick to the same tens.*

3. Apabila bilangan jari menjadi sifar, tukarkan kepada nilai sepuluh di tangan kanan / *When the number of fingers become zero, change to ten on right hand.*

Penerangan untuk kod warna Jadual Sifir /

Explanation on the colour coding for the Times Table

Selain dari menghafal menggunakan jari, anda juga boleh menghafal menggunakan kod warna pada Jadual Sifir…

Besides memorizing using fingers, you can also memorize using colour coding in the Times Tables…

Nombor ganjil: tulisan hitam	Sifar & nombor genap: tulisan putih
Odd numbers: in black font	Zero & even numbers: in white font

Nombor putih (sa) x nombor putih (sa) = nombor putih (sa)
Nombor putih (sa) x nombor hitam (sa) = nombor putih (sa)
Nombor hitam (sa) x nombor hitam (sa) = nombor hitam (sa)

White number (ones) x white number (ones) = white number (ones)
White number (ones) x black number (ones) = white number (ones)
Black number (ones) x black number (ones) = black number (ones)

Sifar di hadapan nombor 1 hingga 9 tiada nilai.
Sifar di belakang nombor 1 hingga 9 mempunyai nilai.

Zero in front of number 1 until 9 has no values.
Zero behind the number 1 until 9 has values.

Kod warna hampir mengikut turutan warna dalam pelangi:

Colour coding almost following the sequence in the rainbow:

0 - white font, black box

1 - black font, pink box

2 - white font, red box

3 - black font, yellow box

4 - white font, orange box

5 - black font, light green box

6 - white font, bright green box

7 - black font, light blue box

8 - white font, bright blue box

9 - black font, light purple box

| 0 |
| 1 |
| 2 |
| 3 |
| 4 |
| 5 |
| 6 |
| 7 |
| 8 |
| 9 |

Sifir 1 / Times 1

(mengikut kotak berwarna kuning di muka surat 1 hingga 2 / following yellow boxes on page 1 to 2)

Sifir 1 adalah yang paling senang.

Times 1 is the easiest.

1

Cuma senaraikan ini.

Just list down these.

2

Kemudian, tuliskan ini.

Then, write down these.

1	×	1	=
2	×	1	=
3	×	1	=
4	×	1	=
5	×	1	=
6	×	1	=
7	×	1	=
8	×	1	=
9	×	1	=
10	×	1	=
11	×	1	=
12	×	1	=

1
2
3
4
5
6
7
8
9
10
11
12

Ini adalah Sifir 1 mengikut kod warna.

This is the Times 1 according to the colour coding.

0	1	×	0	1	=	0	0	1
0	2	×	0	1	=	0	0	2
0	3	×	0	1	=	0	0	3
0	4	×	0	1	=	0	0	4
0	5	×	0	1	=	0	0	5
0	6	×	0	1	=	0	0	6
0	7	×	0	1	=	0	0	7
0	8	×	0	1	=	0	0	8
0	9	×	0	1	=	0	0	9
1	0	×	0	1	=	0	1	0
1	1	×	0	1	=	0	1	1
1	2	×	0	1	=	0	1	2

Sekarang, kita sudah menghafal setakat ini (petak berwarna) …

Now, we have memorized these so far (coloured boxes) …

1	2	3	4	5	6	7	8	9	10	11	12
2	4	6	8	10	12	14	16	18	20	22	24
3	6	9	12	15	18	21	24	27	30	33	36
4	8	12	16	20	24	28	32	36	40	44	48
5	10	15	20	25	30	35	40	45	50	55	60
6	12	18	24	30	36	42	48	54	60	66	72
7	14	21	28	35	42	49	56	63	70	77	84
8	16	24	32	40	48	56	64	72	80	88	96
9	18	27	36	45	54	63	72	81	90	99	108
10	20	30	40	50	60	70	80	90	100	110	120
11	22	33	44	55	66	77	88	99	110	121	132
12	24	36	48	60	72	84	96	108	120	132	144

DONE

SIFIR 10 / TIMES 10

(mengikut kotak berwarna biru di muka surat 2 hingga 4 / following blue boxes on page 2 to 4)

1. Tuliskan nombor 1 hingga 10.

   Write down numbers 1 until 10.

2. Tambahkan sifar (0) di penghujung setiap nombor.

   Add zero (0) at the end of each number.

3. Kemudian, tuliskan ini.

   Then, write down these.

1	×	10	=	1	0
2	×	10	=	2	0
3	×	10	=	3	0
4	×	10	=	4	0
5	×	10	=	5	0
6	×	10	=	6	0
7	×	10	=	7	0
8	×	10	=	8	0
9	×	10	=	9	0
10	×	10	=	10	0
11	×	10	=	11	0
12	×	10	=	12	0

Ini adalah Sifir 10 mengikut kod warna.

This is the Times 10 according to the colour coding.

0	1	×	1	0	=	0	1	0
0	2	×	1	0	=	0	2	0
0	3	×	1	0	=	0	3	0
0	4	×	1	0	=	0	4	0
0	5	×	1	0	=	0	5	0
0	6	×	1	0	=	0	6	0
0	7	×	1	0	=	0	7	0
0	8	×	1	0	=	0	8	0
0	9	×	1	0	=	0	9	0
1	0	×	1	0	=	1	0	0
1	1	×	1	0	=	1	1	0
1	2	×	1	0	=	1	2	0

Sekarang, kita sudah menghafal setakat ini (petak berwarna) ...

Now, we have memorized these so far (coloured boxes) ...

1	2	3	4	5	6	7	8	9	10	11	12
2	4	6	8	10	12	14	16	18	20	22	24
3	6	9	12	15	18	21	24	27	30	33	36
4	8	12	16	20	24	28	32	36	40	44	48
5	10	15	20	25	30	35	40	45	50	55	60
6	12	18	24	30	36	42	48	54	60	66	72
7	14	21	28	35	42	49	56	63	70	77	84
8	16	24	32	40	48	56	64	72	80	88	96
9	18	27	36	45	54	63	72	81	90	99	108
10	20	30	40	50	60	70	80	90	100	110	120
11	22	33	44	55	66	77	88	99	110	121	132
12	24	36	48	60	72	84	96	108	120	132	144

✔ DONE 👍

STAIL SIFIR 5 / TIMES 5 STYLE

(gunakan tangan kanan dan kiri /use right and left hands)

(tangan kanan: + 5 jari;
bila capai nilai 10 di tangan kanan, ganti dengan
tangan kiri: +1 jari & tangan kanan = 0 /

left hand: + 5 fingers;
when counts reach 10 on right hand, substitute with
left hand: + 1 finger & right hand = 0)

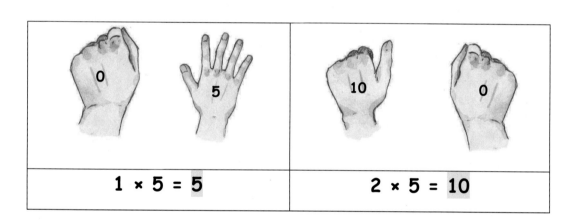

1 × 5 = 5

2 × 5 = 10

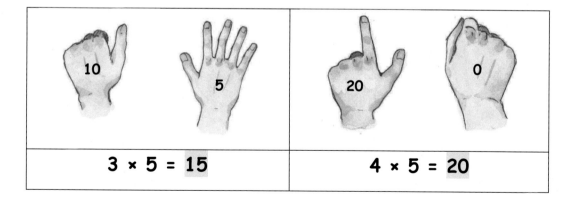

3 × 5 = 15

4 × 5 = 20

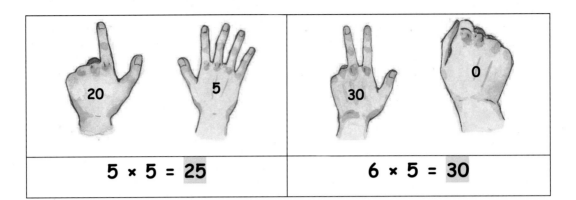

| 5 × 5 = 25 | 6 × 5 = 30 |

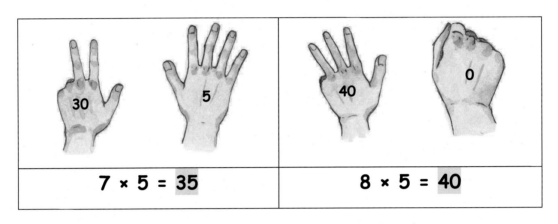

| 7 × 5 = 35 | 8 × 5 = 40 |

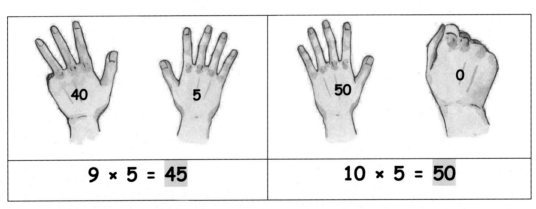

| 9 × 5 = 45 | 10 × 5 = 50 |

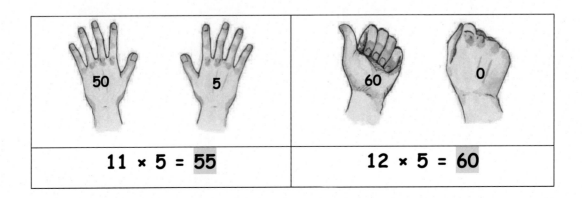

| 11 × 5 = 55 | 12 × 5 = 60 |

Ini adalah Sifir 5 mengikut kod warna.

This is the Times 5 according to the colour coding.

0	1	×	0	5	=	0	0	5
0	2	×	0	5	=	0	1	0
0	3	×	0	5	=	0	1	5
0	4	×	0	5	=	0	2	0
0	5	×	0	5	=	0	2	5
0	6	×	0	5	=	0	3	0
0	7	×	0	5	=	0	3	5
0	8	×	0	5	=	0	4	0
0	9	×	0	5	=	0	4	5
1	0	×	0	5	=	0	5	0
1	1	×	0	5	=	0	5	5
1	2	×	0	5	=	0	6	0

Sekarang, kita sudah menghafal setakat ini (petak berwarna) ...

Now, we have memorized these so far (coloured boxes) ...

1	2	3	4	5	6	7	8	9	10	11	12
2	4	6	8	10	12	14	16	18	20	22	24
3	6	9	12	15	18	21	24	27	30	33	36
4	8	12	16	20	24	28	32	36	40	44	48
5	10	15	20	25	30	35	40	45	50	55	60
6	12	18	24	30	36	42	48	54	60	66	72
7	14	21	28	35	42	49	56	63	70	77	84
8	16	24	32	40	48	56	64	72	80	88	96
9	18	27	36	45	54	63	72	81	90	99	108
10	20	30	40	50	60	70	80	90	100	110	120
11	22	33	44	55	66	77	88	99	110	121	132
12	24	36	48	60	72	84	96	108	120	132	144

DONE

STAIL SIFIR 11 / *TIMES 11 STYLE*

(gunakan tangan kanan dan kiri /use right and left hands)

(tangan kiri: + 1 jari, tangan kanan: + 1 jari / left hand: + 1 finger, right hand: + 1 finger)

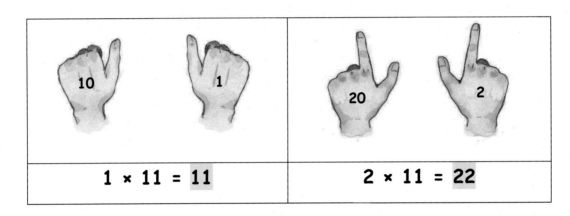

| 1 × 11 = 11 | 2 × 11 = 22 |

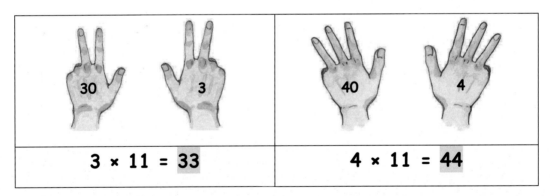

| 3 × 11 = 33 | 4 × 11 = 44 |

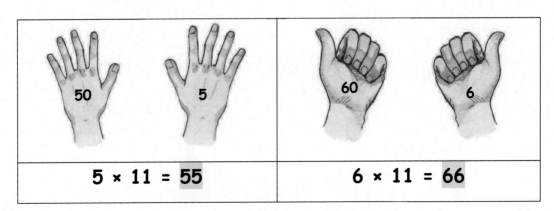

| 5 × 11 = 55 | 6 × 11 = 66 |

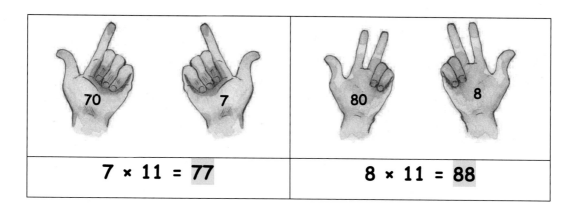

7 × 11 = **77**

8 × 11 = **88**

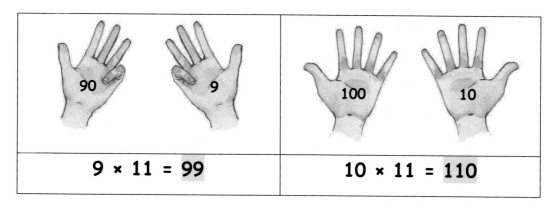

9 × 11 = **99**

10 × 11 = **110**

Selepas kiraan 110, anda perlu menggunakan satu kali tangan kiri dan dua kali tangan kanan:

After any count more than 110, you ought to use one time of left hand and two times of right hand:

Left hand		
Right hand		
	11 × 11 = 121	12 × 11 = 132

Ini adalah Sifir 11 mengikut kod warna.

This is the Times 11 according to the colour coding.

0	1	×	1	1	=	0	1	1
0	2	×	1	1	=	0	2	2
0	3	×	1	1	=	0	3	3
0	4	×	1	1	=	0	4	4
0	5	×	1	1	=	0	5	5
0	6	×	1	1	=	0	6	6
0	7	×	1	1	=	0	7	7
0	8	×	1	1	=	0	8	8
0	9	×	1	1	=	0	9	9
1	0	×	1	1	=	1	1	0
1	1	×	1	1	=	1	2	1
1	2	×	1	1	=	1	3	2

Sekarang, kita sudah menghafal setakat ini (petak berwarna) ...

Now, we have memorized these so far (coloured boxes) ...

1	2	3	4	5	6	7	8	9	10	11	12
2	4	6	8	10	12	14	16	18	20	22	24
3	6	9	12	15	18	21	24	27	30	33	36
4	8	12	16	20	24	28	32	36	40	44	48
5	10	15	20	25	30	35	40	45	50	55	60
6	12	18	24	30	36	42	48	54	60	66	72
7	14	21	28	35	42	49	56	63	70	77	84
8	16	24	32	40	48	56	64	72	80	88	96
9	18	27	36	45	54	63	72	81	90	99	108
10	20	30	40	50	60	70	80	90	100	110	120
11	22	33	44	55	66	77	88	99	110	121	132
12	24	36	48	60	72	84	96	108	120	132	144

✔ DONE 👍

STAIL SIFIR 2 / TIMES 2 STYLE

(Stail senyap & sebut ATAU tangan kanan: + 2 jari / Silent & chant style OR right hand: + 2 fingers)

a) Stail senyap & sebut / Silent & chant style:

1	Senyap / Silent
2	Sebut / Chant
3	Senyap / Silent
4	Sebut / Chant
5	Senyap / Silent
6	Sebut / Chant
7	Senyap / Silent
8	Sebut / Chant
9	Senyap / Silent
10	Sebut / Chant
11	Senyap / Silent
12	Sebut / Chant

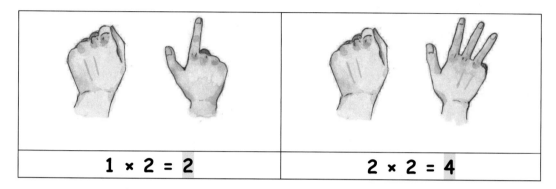 Scan me

b) Tangan kanan: + 2 jari/right hand: + 2 fingers:

(gunakan tangan kanan dan kiri /use right and left hands)

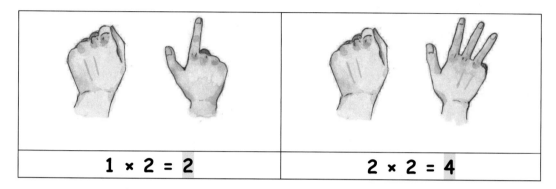

1 × 2 = 2	2 × 2 = 4

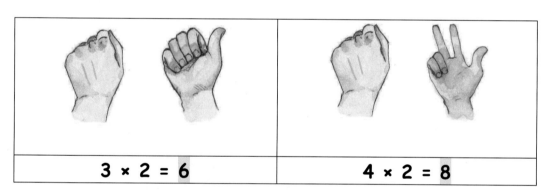

3 × 2 = 6	4 × 2 = 8

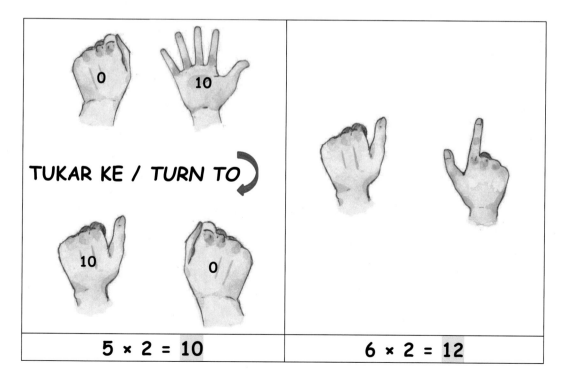

TUKAR KE / TURN TO

5 × 2 = 10	6 × 2 = 12

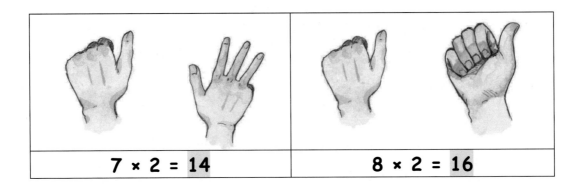

7 × 2 = 14   8 × 2 = 16

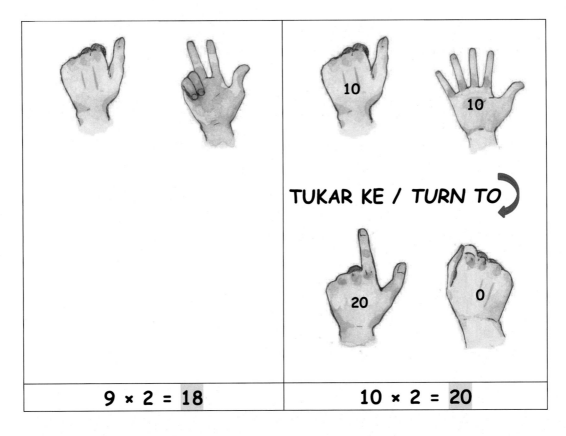

TUKAR KE / TURN TO

9 × 2 = 18   10 × 2 = 20

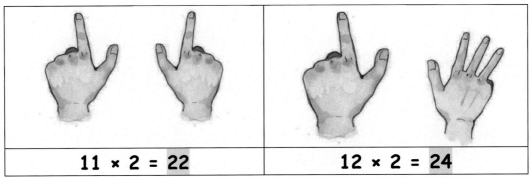

11 × 2 = 22   12 × 2 = 24

Ini adalah Sifir 2 mengikut kod warna.

This is the Times 2 according to the colour coding.

0	1	×	0	2	=	0	0	2
0	2	×	0	2	=	0	0	4
0	3	×	0	2	=	0	0	6
0	4	×	0	2	=	0	0	8
0	5	×	0	2	=	0	1	0
0	6	×	0	2	=	0	1	2
0	7	×	0	2	=	0	1	4
0	8	×	0	2	=	0	1	6
0	9	×	0	2	=	0	1	8
1	0	×	0	2	=	0	2	0
1	1	×	0	2	=	0	2	2
1	2	×	0	2	=	0	2	4

Sekarang, kita sudah menghafal setakat ini (kotak berwarna) ...

Now, we have memorized these so far (coloured boxes) ...

1	2	3	4	5	6	7	8	9	10	11	12
2	4	6	8	10	12	14	16	18	20	22	24
3	6	9	12	15	18	21	24	27	30	33	36
4	8	12	16	20	24	28	32	36	40	44	48
5	10	15	20	25	30	35	40	45	50	55	60
6	12	18	24	30	36	42	48	54	60	66	72
7	14	21	28	35	42	49	56	63	70	77	84
8	16	24	32	40	48	56	64	72	80	88	96
9	18	27	36	45	54	63	72	81	90	99	108
10	20	30	40	50	60	70	80	90	100	110	120
11	22	33	44	55	66	77	88	99	110	121	132
12	24	36	48	60	72	84	96	108	120	132	144

✔ DONE 👍

| UNIT 8 | **STAIL SIFIR 12 / _TIMES 12 STYLE_** |

(gunakan tangan kanan dan kiri /_use right and left hands_)

(tangan kiri: + 1 jari, tangan kanan: + 2 jari / _left hand: + 1 finger, right hand: + 2 fingers_)

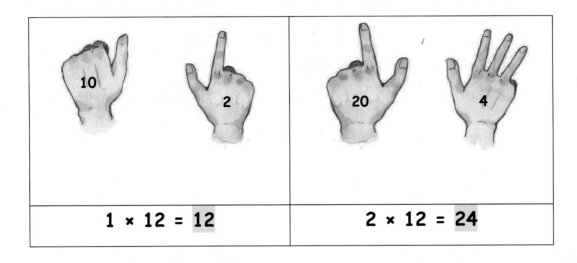

| 1 × 12 = 12 | 2 × 12 = 24 |

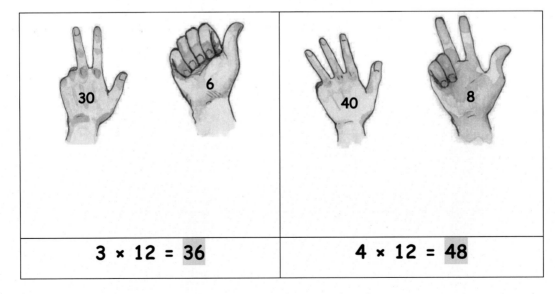

| 3 × 12 = 36 | 4 × 12 = 48 |

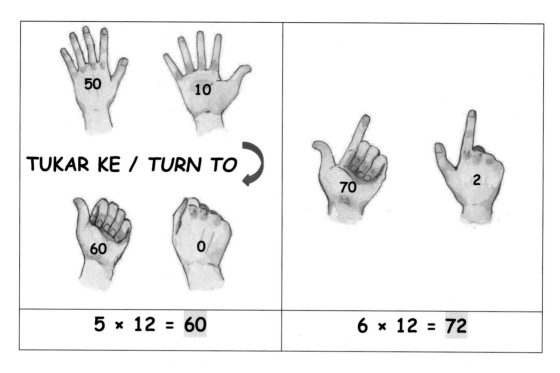

TUKAR KE / TURN TO

5 × 12 = 60

6 × 12 = 72

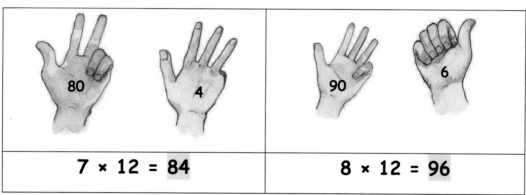

7 × 12 = 84

8 × 12 = 96

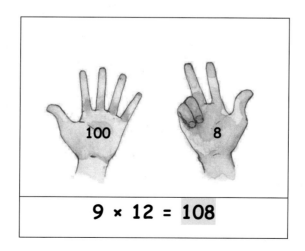

9 × 12 = 108

Selepas kiraan 110, anda perlu menggunakan satu tangan kiri dan dua tangan kanan:

After any count more than 110, you ought to use one time of left hand and two times of right hand:

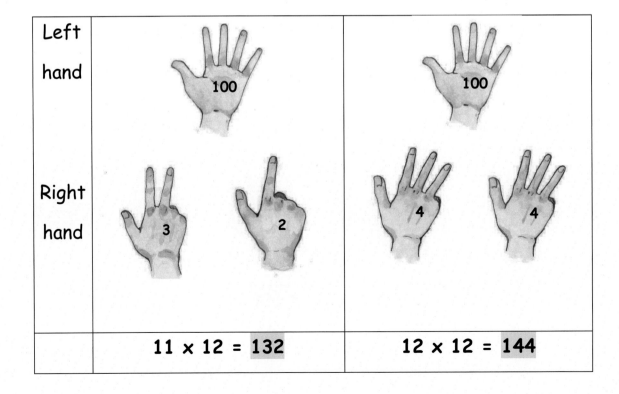

Left hand	100
Right hand	2 0
	10 x 12 = 120

Left hand	100	100
Right hand	3 2	4 4
	11 x 12 = 132	**12 x 12 = 144**

Ini adalah Sifir 12 mengikut kod warna.

This is the Times 12 according to the colour coding.

0	1	×	1	2	=	0	1	2
0	2	×	1	2	=	0	2	4
0	3	×	1	2	=	0	3	6
0	4	×	1	2	=	0	4	8
0	5	×	1	2	=	0	6	0
0	6	×	1	2	=	0	7	2
0	7	×	1	2	=	0	8	4
0	8	×	1	2	=	0	9	6
0	9	×	1	2	=	1	0	8
1	0	×	1	2	=	1	2	0
1	1	×	1	2	=	1	3	2
1	2	×	1	2	=	1	4	4

Sekarang, kita sudah menghafal setakat ini (kotak berwarna) ...

Now, we have memorized these so far (coloured boxes) ...

1	2	3	4	5	6	7	8	9	10	11	12
2	4	6	8	10	12	14	16	18	20	22	24
3	6	9	12	15	18	21	24	27	30	33	36
4	8	12	16	20	24	28	32	36	40	44	48
5	10	15	20	25	30	35	40	45	50	55	60
6	12	18	24	30	36	42	48	54	60	66	72
7	14	21	28	35	42	49	56	63	70	77	84
8	16	24	32	40	48	56	64	72	80	88	96
9	18	27	36	45	54	63	72	81	90	99	108
10	20	30	40	50	60	70	80	90	100	110	120
11	22	33	44	55	66	77	88	99	110	121	132
12	24	36	48	60	72	84	96	108	120	132	144

✔ DONE 👍

UNIT 9

STAIL SIFIR 9 / TIMES 9 STYLE

(gunakan tangan kanan sahaja /use right hand only)

(- 1 jari / - 1 finger)

Apabila bilangan jari berkurangan, tambahkan satu nilai puluh / When the number of fingers is decreasing, add one tens.

Apabila bilangan jari bertambah, tetap pada nilai puluh yang sama / When the number of fingers is increasing, stick to the same tens.

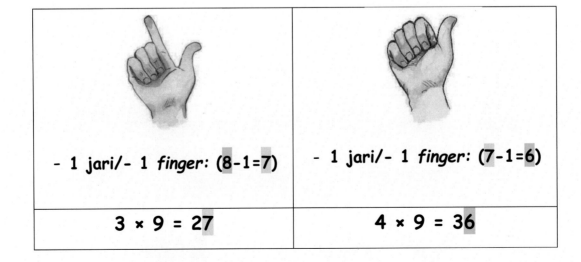

	- 1 jari/- 1 finger: (9-1=8)
1 × 9 = 9	2 × 9 = 18
- 1 jari/- 1 finger: (8-1=7)	- 1 jari/- 1 finger: (7-1=6)
3 × 9 = 27	4 × 9 = 36

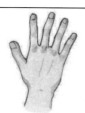

- 1 jari/- 1 *finger*: (6-1=5)

5 × 9 = 45

- 1 jari/- 1 *finger*: (5-1=4)

6 × 9 = 54

- 1 jari/- 1 *finger*: (4-1=3)

7 × 9 = 63

- 1 jari/- 1 *finger*: (3-1=2)

8 × 9 = 72

- 1 jari/- 1 *finger*: (2-1=1)

9 × 9 = 81

- 1 jari/- 1 *finger*: (1-1=0),
tukar ke/*turn to* 10)

10 × 9 = 90

- 1 jari/- 1 *finger*: (10-1=9)

11 × 9 = 99

- 1 jari/- 1 *finger*: (9-1=8)

12 × 9 = 108

Ini adalah Sifir 9 mengikut kod warna.

This is the Times 9 according to the colour coding.

0	1	×	0	9	=	0	0	9
0	2	×	0	9	=	0	1	8
0	3	×	0	9	=	0	2	7
0	4	×	0	9	=	0	3	6
0	5	×	0	9	=	0	4	5
0	6	×	0	9	=	0	5	4
0	7	×	0	9	=	0	6	3
0	8	×	0	9	=	0	7	2
0	9	×	0	9	=	0	8	1
1	0	×	0	9	=	0	9	0
1	1	×	0	9	=	0	9	9
1	2	×	0	9	=	1	0	8

Sekarang, kita sudah menghafal setakat ini (petak berwarna) ...

Now, we have memorized these so far (coloured boxes) ...

1	2	3	4	5	6	7	8	9	10	11	12
2	4	6	8	10	12	14	16	18	20	22	24
3	6	9	12	15	18	21	24	27	30	33	36
4	8	12	16	20	24	28	32	36	40	44	48
5	10	15	20	25	30	35	40	45	50	55	60
6	12	18	24	30	36	42	48	54	60	66	72
7	14	21	28	35	42	49	56	63	70	77	84
8	16	24	32	40	48	56	64	72	80	88	96
9	18	27	36	45	54	63	72	81	90	99	108
10	20	30	40	50	60	70	80	90	100	110	120
11	22	33	44	55	66	77	88	99	110	121	132
12	24	36	48	60	72	84	96	108	120	132	144

✔ DONE 👍

STAIL SIFIR 8 / TIMES 8 STYLE

(gunakan tangan kanan sahaja / use right hand only)

(- 2 jari/- 2 fingers)

Scan me

Apabila bilangan jari berkurangan, tambahkan satu nilai puluh / When the number of fingers is decreasing, add one tens.

Apabila bilangan jari bertambah, tetap pada nilai puluh yang sama / When the number of fingers is increasing, stick to the same tens.

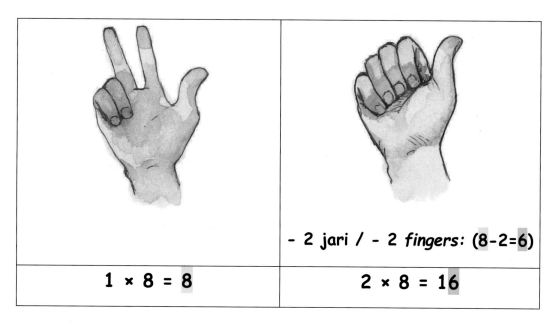

	- 2 jari / - 2 fingers: (8-2=6)
1 × 8 = 8	2 × 8 = 16

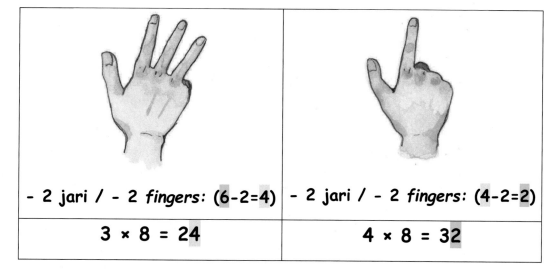

- 2 jari / - 2 fingers: (6-2=4)	- 2 jari / - 2 fingers: (4-2=2)
3 × 8 = 24	4 × 8 = 32

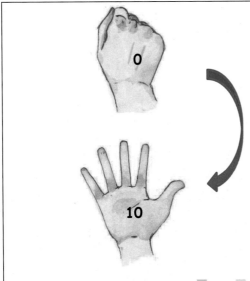

- 2 jari /- 2 fingers: (2-2=0,

tukar ke/turn to 10)

5 × 8 = 40

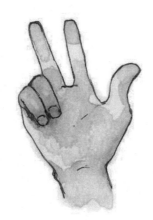

- 2 jari/- 2 fingers: (10-2=8)

6 × 8 = 48

- 2 jari / - 2 fingers: (8-2=6)

7 × 8 = 56

- 2 jari / - 2 fingers: (6-2=4)

8 × 8 = 64

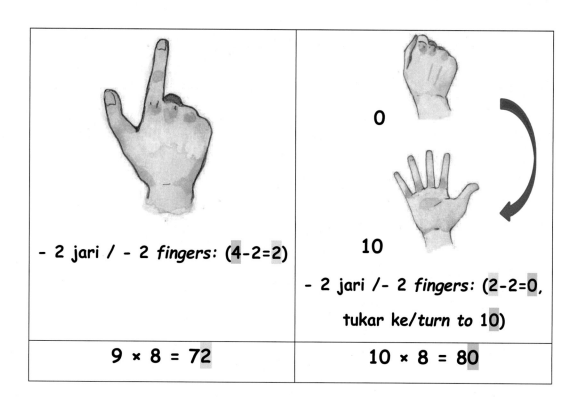

- 2 jari / - 2 fingers: (4-2=2)	- 2 jari /- 2 fingers: (2-2=0, tukar ke/turn to 10)
9 × 8 = 72	10 × 8 = 80

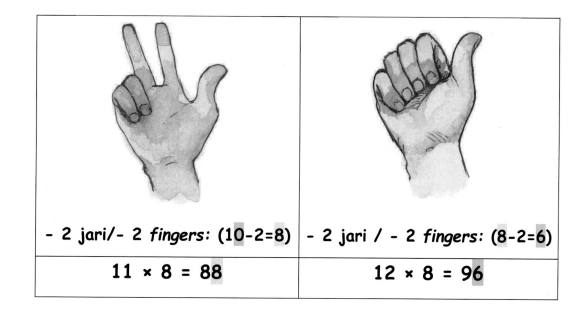

- 2 jari/- 2 fingers: (10-2=8)	- 2 jari / - 2 fingers: (8-2=6)
11 × 8 = 88	12 × 8 = 96

Ini adalah Sifir 8 mengikut kod warna.

This is the Times 8 according to the colour coding.

0	1	×	0	8	=	0	0	8
0	2	×	0	8	=	0	1	6
0	3	×	0	8	=	0	2	4
0	4	×	0	8	=	0	3	2
0	5	×	0	8	=	0	4	0
0	6	×	0	8	=	0	4	8
0	7	×	0	8	=	0	5	6
0	8	×	0	8	=	0	6	4
0	9	×	0	8	=	0	7	2
1	0	×	0	8	=	0	8	0
1	1	×	0	8	=	0	8	8
1	2	×	0	8	=	0	9	6

Sekarang, kita sudah menghafal setakat ini...

Now, we have memorized these so far...

1	2	3	4	5	6	7	8	9	10	11	12
2	4	6	8	10	12	14	16	18	20	22	24
3	6	9	12	15	18	21	24	27	30	33	36
4	8	12	16	20	24	28	32	36	40	44	48
5	10	15	20	25	30	35	40	45	50	55	60
6	12	18	24	30	36	42	48	54	60	66	72
7	14	21	28	35	42	49	56	63	70	77	84
8	16	24	32	40	48	56	64	72	80	88	96
9	18	27	36	45	54	63	72	81	90	99	108
10	20	30	40	50	60	70	80	90	100	110	120
11	22	33	44	55	66	77	88	99	110	121	132
12	24	36	48	60	72	84	96	108	120	132	144

✔ DONE 👍

STAIL SIFIR 6 / TIMES 6 STYLE

(gunakan tangan kanan sahaja / use right hand only)

(+ 1 jari & terbalikkan tangan/ + 1 finger & turn hand)

Apabila bilangan jari berkurangan, tambahkan satu nilai puluh / When the number of fingers is decreasing, add one tens.

Apabila bilangan jari bertambah, tetap pada nilai puluh yang sama / When the number of fingers is increasing, stick to the same tens.

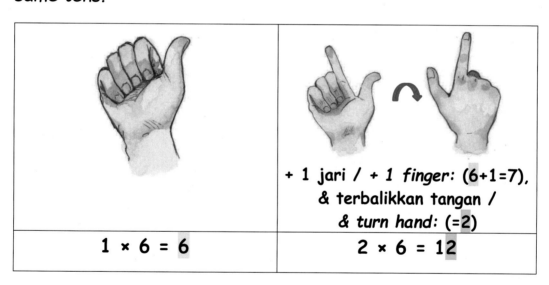

	+ 1 jari / + 1 finger: (6+1=7), & terbalikkan tangan / & turn hand: (=2)
1 × 6 = 6	**2 × 6 = 12**

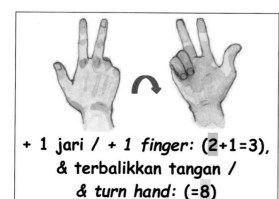
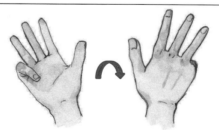

+ 1 jari / + 1 finger: (2+1=3), & terbalikkan tangan / & turn hand: (=8)	+ 1 jari / + 1 finger: (8+1=9), & terbalikkan tangan / & turn hand: (=4)
3 × 6 = 18	**4 × 6 = 24**

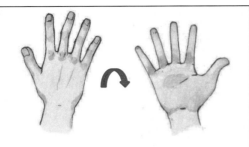

+ 1 jari / + 1 finger: (4+1=5),
& terbalikkan tangan /
& turn hand:
(=10, ambil/take 0)

5 × 6 = 30

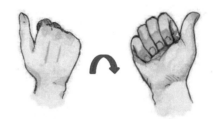

+ 1 jari/+ 1 finger: (0+1=1),
& terbalikkan tangan /
& turn hand: (=6)

6 × 6 = 36

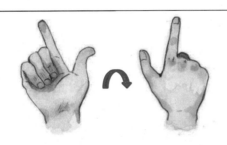

+ 1 jari / + 1 finger: (6+1=7),
& terbalikkan tangan /
& turn hand: (=2)

7 × 6 = 42

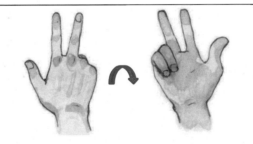

+ 1 jari / + 1 finger: (2+1=3),
& terbalikkan tangan /
& turn hand: (=8)

8 × 6 = 48

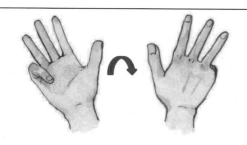

+ 1 jari / + 1 finger: (8+1=9),
& terbalikkan tangan /
& turn hand: (=4)

9 × 6 = 54

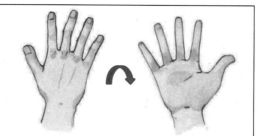

+ 1 jari / + 1 finger: (4+1=5),
& terbalikkan tangan /
& turn hand:
(=10, ambil/take 0)

10 × 6 = 60

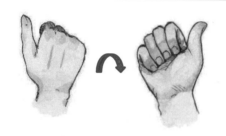

+ 1 jari / + 1 finger: (0+1=1),
& terbalikkan tangan /
& turn hand: (=6)

11 × 6 = 66

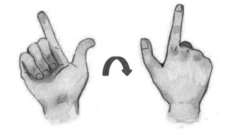

+ 1 jari / + 1 finger: (6+1=7),
& terbalikkan tangan /
& turn hand: (=2)

12 × 6 = 72

Ini adalah Sifir 6 mengikut kod warna.

This is the Times 6 according to the colour coding.

0	1	×	0	6	=	0	0	6
0	2	×	0	6	=	0	1	2
0	3	×	0	6	=	0	1	8
0	4	×	0	6	=	0	2	4
0	5	×	0	6	=	0	3	0
0	6	×	0	6	=	0	3	6
0	7	×	0	6	=	0	4	2
0	8	×	0	6	=	0	4	8
0	9	×	0	6	=	0	5	4
1	0	×	0	6	=	0	6	0
1	1	×	0	6	=	0	6	6
1	2	×	0	6	=	0	7	2

Sekarang, kita sudah menghafal setakat ini...

Now, we have memorized these so far...

1	2	3	4	5	6	7	8	9	10	11	12
2	4	6	8	10	12	14	16	18	20	22	24
3	6	9	12	15	18	21	24	27	30	33	36
4	8	12	16	20	24	28	32	36	40	44	48
5	10	15	20	25	30	35	40	45	50	55	60
6	12	18	24	30	36	42	48	54	60	66	72
7	14	21	28	35	42	49	56	63	70	77	84
8	16	24	32	40	48	56	64	72	80	88	96
9	18	27	36	45	54	63	72	81	90	99	108
10	20	30	40	50	60	70	80	90	100	110	120
11	22	33	44	55	66	77	88	99	110	121	132
12	24	36	48	60	72	84	96	108	120	132	144

✔ DONE 👍

STAIL SIFIR 7 / TIMES 7 STYLE

(gunakan tangan kanan sahaja / use right hand only)

(+ 2 jari & terbalikkan tangan/+ 2 fingers & turn hand)

Apabila bilangan jari berkurangan, tambahkan satu nilai puluh / *When the number of fingers is decreasing, add one tens.*

Apabila bilangan jari bertambah, tetap pada nilai puluh yang sama / *When the number of fingers is increasing, stick to the same tens.*

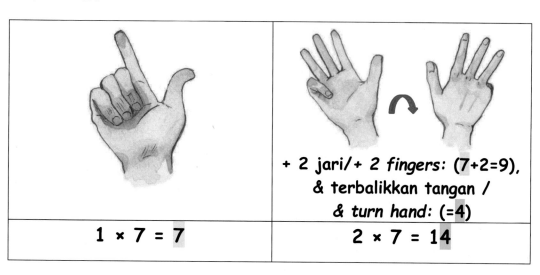

	+ 2 jari/+ 2 fingers: (7+2=9), & terbalikkan tangan / & turn hand: (=4)
1 × 7 = 7	**2 × 7 = 14**

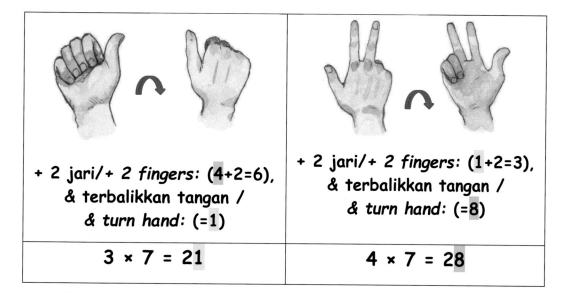

+ 2 jari/+ 2 fingers: (4+2=6), & terbalikkan tangan / & turn hand: (=1)	+ 2 jari/+ 2 fingers: (1+2=3), & terbalikkan tangan / & turn hand: (=8)
3 × 7 = 21	**4 × 7 = 28**

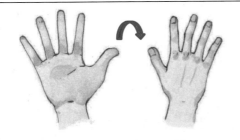

+ 2 jari/+ 2 *fingers*:(8+2=10),
& terbalikkan tangan /
& *turn hand*: (=5)

5 × 7 = 35

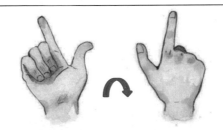

+ 2 jari/+ 2 *fingers*: (5+2=7),
& terbalikkan tangan /
& *turn hand*: (=2)

6 × 7 = 42

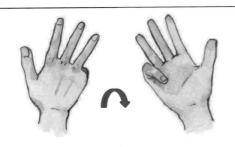

+ 2 jari/+ 2 *fingers*: (2+2=4),
& terbalikkan tangan /
& *turn hand*: (=9)

7 × 7 = 49

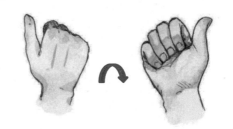

+ 2 jari/+ 2 *fingers*:(9+2=11),
ambil/*take* 1),
& terbalikkan tangan /
& *turn hand*: (=6)

8 × 7 = 56

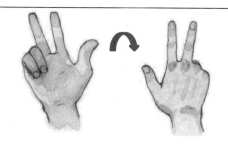 + 2 jari/+ 2 fingers: (6+2=8), & terbalikkan tangan / & turn hand: (=3)	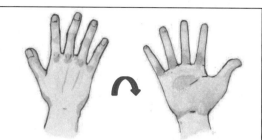 + 2 jari/+ 2 fingers: (3+2=5), & terbalikkan tangan / &turn hand:(=10, ambil/take 0)
9 × 7 = 63	10 × 7 = 70
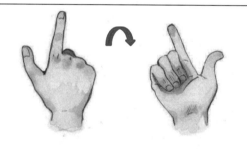 + 2 jari/+ 2 fingers: (0+2=2), & terbalikkan tangan / & turn hand: (=7)	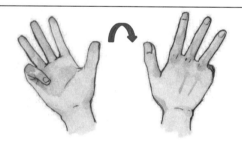 + 2 jari/+ 2 fingers: (7+2=9), & terbalikkan tangan / & turn hand: (=4)
11 × 7 = 77	12 × 7 = 84

Ini adalah Sifir 7 mengikut kod warna.

This is the Times 7 according to the colour coding.

0	1	×	0	7	=	0	0	7
0	2	×	0	7	=	0	1	4
0	3	×	0	7	=	0	2	1
0	4	×	0	7	=	0	2	8
0	5	×	0	7	=	0	3	5
0	6	×	0	7	=	0	4	2
0	7	×	0	7	=	0	4	9
0	8	×	0	7	=	0	5	6
0	9	×	0	7	=	0	6	3
1	0	×	0	7	=	0	7	0
1	1	×	0	7	=	0	7	7
1	2	×	0	7	=	0	8	4

Sekarang, kita sudah menghafal setakat ini...

Now, we have memorized these so far...

1	2	3	4	5	6	7	8	9	10	11	12
2	4	6	8	10	12	14	16	18	20	22	24
3	6	9	12	15	18	21	24	27	30	33	36
4	8	12	16	20	24	28	32	36	40	44	48
5	10	15	20	25	30	35	40	45	50	55	60
6	12	18	24	30	36	42	48	54	60	66	72
7	14	21	28	35	42	49	56	63	70	77	84
8	16	24	32	40	48	56	64	72	80	88	96
9	18	27	36	45	54	63	72	81	90	99	108
10	20	30	40	50	60	70	80	90	100	110	120
11	22	33	44	55	66	77	88	99	110	121	132
12	24	36	48	60	72	84	96	108	120	132	144

✔ DONE 👍

STAIL SIFIR 4 / TIMES 4 STYLE

(gunakan tangan kanan sahaja / use right hand only)

(- 1 jari & terbalikkan tangan/ - 1 finger & turn hand)

Apabila bilangan jari berkurangan, tambahkan satu nilai puluh / When the number of fingers is decreasing, add one tens.

Apabila bilangan jari bertambah, tetap pada nilai puluh yang sama / When the number of fingers is increasing, stick to the same tens.

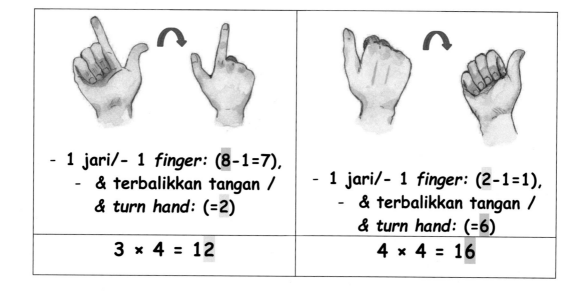

1 × 4 = 4	- 1 jari/- 1 finger: (4-1=3), - & terbalikkan tangan / & turn hand: (=8) 2 × 4 = 8
- 1 jari/- 1 finger: (8-1=7), - & terbalikkan tangan / & turn hand: (=2) 3 × 4 = 12	- 1 jari/- 1 finger: (2-1=1), - & terbalikkan tangan / & turn hand: (=6) 4 × 4 = 16

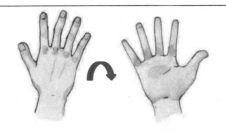

- 1 jari/- 1 *finger*: (6-1=5),
- & terbalikkan tangan /
  & turn hand: (=10)

5 × 4 = 20

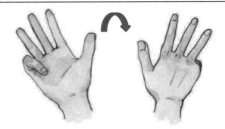

- 1 jari/- 1 *finger*: (10-1=9),
- & terbalikkan tangan /
  & turn hand: (=4)

6 × 4 = 24

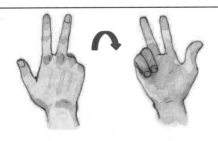

- 1 jari/- 1 *finger*: (4-1=3),
- & terbalikkan tangan /
  & turn hand: (=8)

7 × 4 = 28

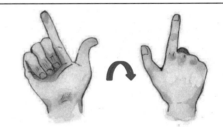

- 1 jari/- 1 *finger*: (8-1=7),
- & terbalikkan tangan /
  & turn hand: (=2)

8 × 4 = 32

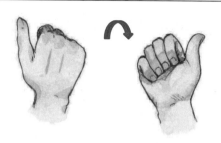

- 1 jari/- 1 *finger*: (2-1=1),
  - & terbalikkan tangan /
  & turn hand: (=6)

9 × 4 = 36

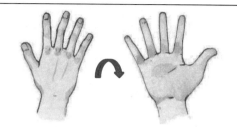

- 1 jari/- 1 *finger*: (6-1=5),
  - & terbalikkan tangan /
  & turn hand: (=10)

10 × 4 = 40

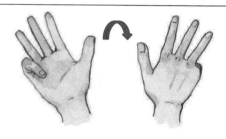

- 1 jari/- 1 *finger*: (10-1=9),
  - & terbalikkan tangan /
  & turn hand: (=4)

11 × 4 = 44

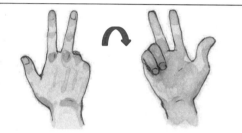

- 1 jari/- 1 *finger*: (4-1=3),
  - & terbalikkan tangan /
  & turn hand: (=8)

12 × 4 = 48

Ini adalah Sifir 4 mengikut kod warna.

This is the Times 4 according to the colour coding.

0	1	×	0	4	=	0	0		4
0	2	×	0	4	=	0	0		8
0	3	×	0	4	=	0		1	2
0	4	×	0	4	=	0		1	6
0	5	×	0	4	=	0		2	0
0	6	×	0	4	=	0		2	4
0	7	×	0	4	=	0		2	8
0	8	×	0	4	=	0		3	2
0	9	×	0	4	=	0		3	6
1	0	×	0	4	=	0		4	0
1	1	×	0	4	=	0		4	4
1	2	×	0	4	=	0		4	8

Sekarang, kita sudah menghafal setakat ini...

Now, we have memorized these so far...

1	2	3	4	5	6	7	8	9	10	11	12
2	4	6	8	10	12	14	16	18	20	22	24
3	6	9	12	15	18	21	24	27	30	33	36
4	8	12	16	20	24	28	32	36	40	44	48
5	10	15	20	25	30	35	40	45	50	55	60
6	12	18	24	30	36	42	48	54	60	66	72
7	14	21	28	35	42	49	56	63	70	77	84
8	16	24	32	40	48	56	64	72	80	88	96
9	18	27	36	45	54	63	72	81	90	99	108
10	20	30	40	50	60	70	80	90	100	110	120
11	22	33	44	55	66	77	88	99	110	121	132
12	24	36	48	60	72	84	96	108	120	132	144

✔ DONE 👍

STAIL SIFIR 3 / TIMES 3 STYLE

(gunakan tangan kanan sahaja / use right hand only)

(- 2 jari & terbalikkan tangan/- 2 fingers & turn hand)

Scan me

Apabila bilangan jari berkurangan, tambahkan satu nilai puluh / *When the number of fingers is decreasing, add one tens.*

Apabila bilangan jari bertambah, tetap pada nilai puluh yang sama / *When the number of fingers is increasing, stick to the same tens.*

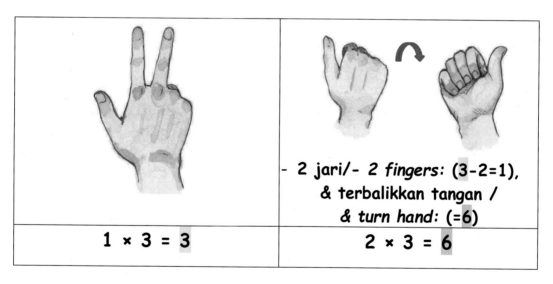

	- 2 jari/- 2 fingers: (3-2=1), & terbalikkan tangan / & turn hand: (=6)
1 × 3 = 3	2 × 3 = 6

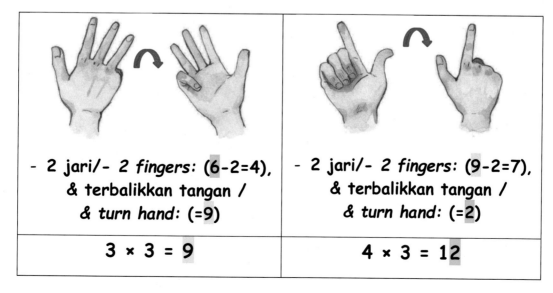

- 2 jari/- 2 fingers: (6-2=4), & terbalikkan tangan / & turn hand: (=9)	- 2 jari/- 2 fingers: (9-2=7), & terbalikkan tangan / & turn hand: (=2)
3 × 3 = 9	4 × 3 = 12

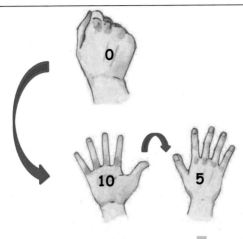

- 2 jari/- 2 fingers: (2-2=0,
  tukar ke/turn to 10),
  & terbalikkan tangan /
  & turn hand: (=5)

5 × 3 = 15

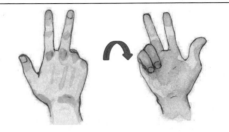

- 2 jari/- 2 fingers:
  (5-2=3),
  & terbalikkan tangan /
  & turn hand: (=8)

6 × 3 = 18

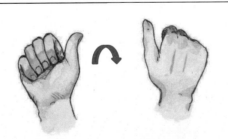

- 2 jari/- 2 fingers: (8-2=6),
  & terbalikkan tangan /
  & turn hand: (=1)

7 × 3 = 21

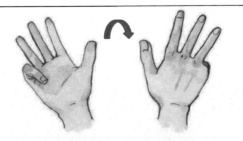

- 2 jari/- 2 fingers:
  (11-2=9),
  & terbalikkan tangan /
  & turn hand: (=4)

8 × 3 = 24

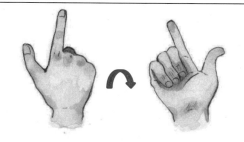

- 2 jari/- 2 fingers: (4-2=2),
& terbalikkan tangan /
& turn hand: (=7)

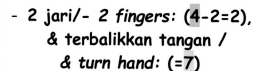

9 × 3 = 27

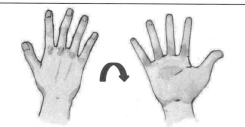

- 2 jari/- 2 fingers: (7-2=5),
& terbalikkan tangan /
& turn hand: (=10)

10 × 3 = 30

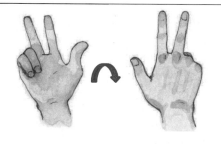

- 2 jari/- 2 fingers:
(10-2=8),
& terbalikkan tangan /
& turn hand: (=3)

11 × 3 = 33

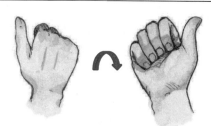

- 2 jari/- 2 fingers: (3-2=1),
& terbalikkan tangan /
& turn hand: (=6)

12 × 3 = 36

Ini adalah Sifir 3 mengikut kod warna.

This is the Times 3 according to the colour coding.

0	1	×	0	3	=	0	0	3
0	2	×	0	3	=	0	0	6
0	3	×	0	3	=	0	0	9
0	4	×	0	3	=	0	1	2
0	5	×	0	3	=	0	1	5
0	6	×	0	3	=	0	1	8
0	7	×	0	3	=	0	2	1
0	8	×	0	3	=	0	2	4
0	9	×	0	3	=	0	2	7
1	0	×	0	3	=	0	3	0
1	1	×	0	3	=	0	3	3
1	2	×	0	3	=	0	3	6

Sekarang, kita sudah menghafal setakat ini…

Now, we have memorized these so far…

1	2	3	4	5	6	7	8	9	10	11	12
2	4	6	8	10	12	14	16	18	20	22	24
3	6	9	12	15	18	21	24	27	30	33	36
4	8	12	16	20	24	28	32	36	40	44	48
5	10	15	20	25	30	35	40	45	50	55	60
6	12	18	24	30	36	42	48	54	60	66	72
7	14	21	28	35	42	49	56	63	70	77	84
8	16	24	32	40	48	56	64	72	80	88	96
9	18	27	36	45	54	63	72	81	90	99	108
10	20	30	40	50	60	70	80	90	100	110	120
11	22	33	44	55	66	77	88	99	110	121	132
12	24	36	48	60	72	84	96	108	120	132	144

✔ DONE 👍

UNIT 15 — PENUTUP / CLOSING

Sekarang, anda telah berjaya menghafal kesemua 12 Jadual Sifir… **Tahniah** diucapkan! Semoga anda dapat menggunakan sifir-sifir 1 hingga 12 ini untuk menjawab soalan matematik darab dan bahagi kemudian nanti.

Now, you have successfully memorized the 12 Times Tables… Congratulations! You may use the 1 to 12 times to answer the multiplication and division maths questions in future.

Terima kasih di atas perhatian yang diberikan

Thank you very much for your kind attention